Yoshinari Minami

Mechanism of GRAVITY Generation

Yoshinari Minami

Mechanism of GRAVITY Generation

-why apples fall-

LAP LAMBERT Academic Publishing

Imprint
Any brand names and product names mentioned in this book are subject to trademark, brand or patent protection and are trademarks or registered trademarks of their respective holders. The use of brand names, product names, common names, trade names, product descriptions etc. even without a particular marking in this work is in no way to be construed to mean that such names may be regarded as unrestricted in respect of trademark and brand protection legislation and could thus be used by anyone.

Cover image: www.ingimage.com

Publisher:
LAP LAMBERT Academic Publishing
is a trademark of
International Book Market Service Ltd., member of OmniScriptum Publishing Group
17 Meldrum Street, Beau Bassin 71504, Mauritius

Printed at: see last page
ISBN: 978-620-2-53059-0

Copyright © Yoshinari Minami
Copyright © 2020 International Book Market Service Ltd., member of OmniScriptum Publishing Group

Acknowledgments

The author wishes to express his sincere thanks to Dr. Giovanni Vulpetti, Dr. Salvatore Santoli and Dr. Claudio Maccone for invaluable discussions, and Paul Murad (CEO Morningstar Applied Physics, LLC; ret. US Department of Defense), Marc G. Millis, Herman David Froning, Dr. Franklin Mead and Prof. Giancarlo Genta for suggesting the importance in this work.

Preface

This is a study of the mechanism of gravity generation.

Why do apples fall?

Gravity was a pushing force, not a pulling force.

Given a priori assumption that space as a vacuum has a physical fine structure like continuum, it enables us to apply a continuum mechanics to the so-called "vacuum" of space.

The apples on the Earth will not be pulled by the Earth and fall, but will be pushed and fall in the direction of the Earth due to the pressure of the field in the curved space area around the Earth.

This book is an attempt to explain the cause of gravity by applying the mechanical structure of space to General Relativity, and is arranged in three chapters as follows.

Chapter 1 (Introduction) explains gravity from the view point of Newtonian mechanics and General Relativity, and introduces the mechanism of gravity newly viewed from the mechanical structure of space.

Chapter 2 is the main part, and introduces the mechanism of gravity with convincing figures along with the numerical values based on concrete formulas.

Chapter 3 introduces the mechanism of gravity described in Chapter 2 in more detail using an example of verification from a multifaceted perspective.

The book concludes that the mass on the Earth will not be pulled by the Earth and fall, but will be pushed and fall in the direction of the Earth due to the pressure of the field in the curved space area around the Earth.

Although the spatial curvature at the surface of the Earth is very small value, i.e., $1.71 \times 10^{-23} (1/m^2)$, it is enough value to produce 1G (9.8 m/s^2) acceleration.

Gravitation can be explained as a pressure field induced by the curvature of space.

Yoshinari Minami

Advanced Space Propulsion Investigation Laboratory (ASPIL)
(Formerly NEC Space Development Division), JAPAN

CONTENTS

CHAPTER 1. Introduction

Release the stone from your hand and it will fall to the ground naturally. The moon orbits the Earth in about a month. The "phenomenon of the moon orbiting the Earth" and the "phenomenon of apples falling on the ground from trees" are actually based on exactly the same principle. Isaac Newton noticed in 1665 that universal gravitation on the ground could be working as well for the moon.

This principle is now called the law of universal gravitation. Newton is said to be inspired by seeing the apple fall from the tree to the ground.

Apples fall from the tree. Why do apples fall?

It has been taught that apples are attracted to the Earth by Newton's universal gravitation, meaning that apples are pulled by the Earth and fall.

The question of why there is gravitation will not give a satisfactory answer to anyone on the Earth.

The first to discover this gravitation was in Newton, England, who, after seeing the apple fall from the tree, reportedly found that the apple was pulled by the Earth and fell, meaning that the Earth had gravitation.

According to Einstein's theory of General Relativity, the following Fig.1.1 is often seen and explained in general books, but it is an unfriendly explanation. After all, it can only be understood that a small object is drawn into the space-time depression around the Earth and falls, and lacks a clear principle of why the small object falls.

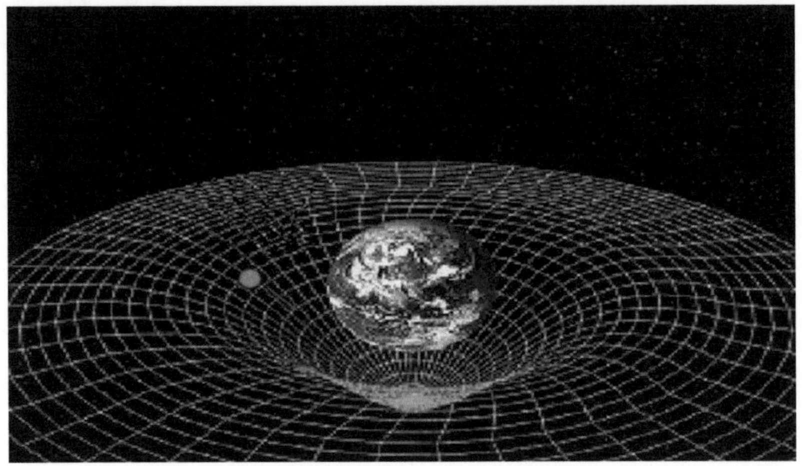

Fig. 1.1. Curved space around the Earth

According to the explanation of Fig. 1.1, an explanatory drawing depicts the state where the Earth is dropped into a two-dimensional grid pattern plane. As can be seen from the explanatory drawing, a state in which the lattice pattern is distorted can be visually recognized, and the distorted lattice pattern itself can be interpreted as gravity. If this illustration is likened to ordinary people, it is the same as a heavy object sinking on a trampoline.

In this figure and description, the flat grid-like space is depressed due to the mass of the Earth, and the space is depressed. It's just a misconception that the small object rolls toward the periphery of the dent, but it is not a consistent explanation of the relationship between the Earth, the curved space, and the falling small object.

In General Relativity, geodesic is a generalization of a "straight line" over a curved space-time. The world line of particles that do not receive any external force other than gravity is a kind of geodesic line and is important. In other words, particles in free motion or free fall move along the geodesic line. However, in general, movement requires force to move. There is no mechanism to explain force generation.

Furthermore, in General Relativity, gravity is thought to be a consequence of the geometry of the curved space-time, not of force, and the source of the space-time curvature is the stress energy tensor (for example, representing matter). Thus, the orbit of a planet orbiting a star is a projection of a geodesic on a curved four-dimensional space-time into three-dimensional space (Fig. 1.2).

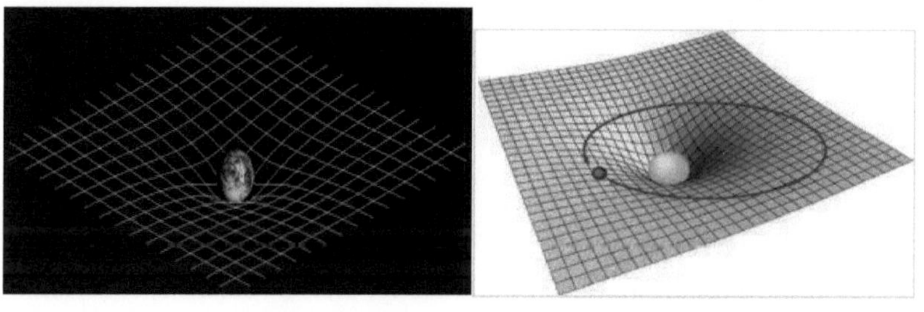

(a) A hollow space around the Earth

(b) An object that falls to the bottom of a hollow space

Fig. 1.2. Explanation of gravity by general relativity

If it is pulled by the Earth, there is no explanation of the mechanism principle. Instead of being pulled by the Earth, the apples are pushed to the Earth by being pushed from the curving space around which the Earth is created.

Gravity was a pushing force, not a pulling force.

The conclusion of General Relativity is that if there is an object mass, the surrounding space will be distorted. The space around a huge mass like the

Earth is curved like a sphere. The pressure in that space is gravity. In other words, gravity is not the "pulling force" of the Earth but the "pushing force" of the space. The space has the property of expanding, contracting, and bending like rubber. If there is a huge object like the Earth, when a human enters the bathtub, the surrounding space will be distorted as if the bath water overflows and it will change its shape.

In general, the mechanism of the pushing phenomenon is clearer than the pulling phenomenon as shown in Fig.1.3.

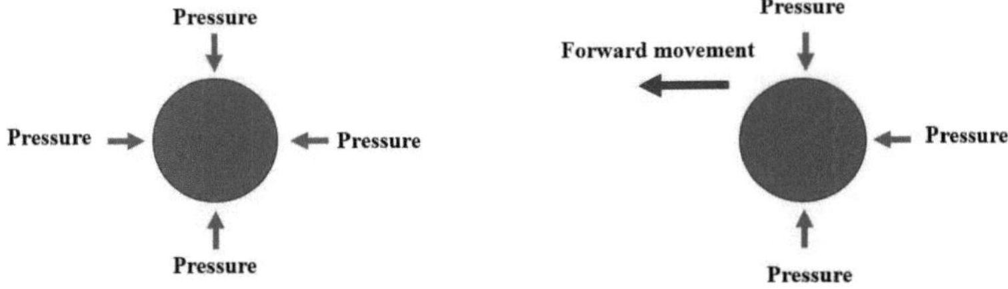

(a) Cannot proceed due to pressure balance.   (b) Pressure balance collapses and is pushed forward.

Fig. 1.3. The ball is pushed forward due to the pressure difference of the ball

If the pressure from the surroundings is evenly and well-balanced to the ball, the ball does not move and stands still (Fig.1.3 (a)). However, if the pressure is out of balance, the ball will move toward a weaker point (Fig.1.3 (b)).

As shown in Fig.1.4, the gravitational field around the Earth is multiply covered by concentric or spherical curved spaces centered on the Earth (see Fig.1.3 (a)).

Considering the case of the Earth, the curvature of space is spherically symmetric about the Earth and is fixed to the Earth, so the Earth itself cannot move due to the curvature of the space generated by the Earth.

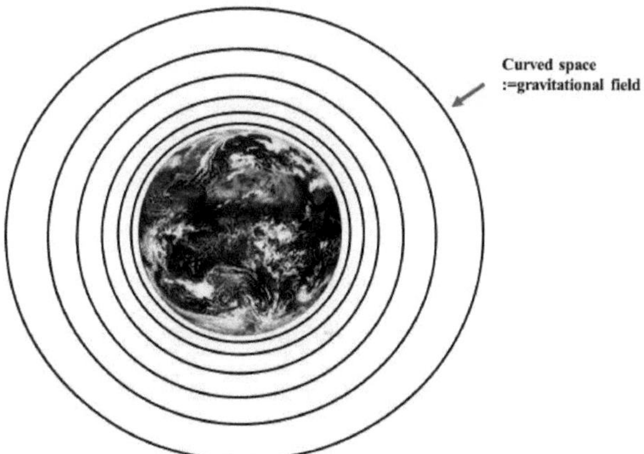

Fig. 1.4. Gravitational field around the Earth is a curved space that is concentric or spherical about the Earth

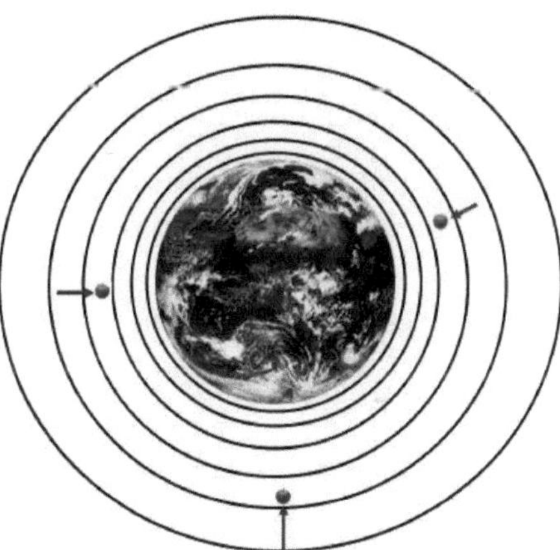

Fig. 1.5. Since the apples on the Earth are independently in the curved spatial regions of the Earth, the apples fall under the pressure generated in the curved spatial regions

However, as shown in Fig.1.5, the apples on the Earth are independently in the curved spatial region of the Earth. Since the apple exists in the curved spatial region from the curved spatial layer at the apple's position to the curved spatial layer at the distant position, the apple is pushed by the generated curved space (i.e., pressure) and falls.

Considering the dynamics of the soap bubble surface, you can see the generation of gravity. The continuum mechanics understands that the surface of the soap bubble is swollen by surface tension, but in fact, a force toward the center of the soap bubble is acting. If this is applied to the space as it is, the space curved in a spherical shape applies pressure toward the center of the sphere. This is the reality of gravity.

Considering this, if the space is artificially curved in a certain area, gravity will be generated there, and the material in the gravitational field should move due to the space pressure. The problem is how to curve the space. Not only the mass but also the energy there can curve the space.

Although the spatial curvature at the surface of the Earth is very small value, i.e., $1.71 \times 10^{-23} (1/m^2)$, it is enough value to produce 1G (9.8 m/s^2) acceleration.

This book is an attempt to explain the cause of gravitation by applying the mechanical structure of space to General Relativity. The concept of mechanical structure of space was primarily intended for application to space propulsion physics. Author has proposed new space propulsion concept at international conferences and peer-reviewed journals since 1988 [1].   However, this book focuses on the cause of gravitational effects.

The mass on the Earth will not be pulled by the Earth and fall, but will be pushed and fall in the direction of the Earth due to the pressure of the field in the curved space area around the Earth.

Dynamic structure of space

Given a priori assumption that space as a vacuum has a physical fine structure like continuum, it enables us to apply a continuum mechanics to the so-called "vacuum" of space. Minami proposed a hypothesis for mechanical property of space-time in 1988 [1]. A primary motive was to research in the realm of space propulsion theory. His propulsion principle using the substantial physical structure of space-time is based on this hypothesis [1-7].

In this book, a fundamental concept of space-time that focuses on theoretically innate properties of space including strain and curvature is described. Assuming that space as vacuum is an infinite continuum, space can be considered as a kind of transparent elastic field. That is, space as a vacuum performs the motions of deformation such as expansion, contraction, elongation, torsion and bending. The latest expanding universe theories (Friedmann, de Sitter, inflationary cosmological model) support this assumption. Space can be regarded as an elastic body like rubber. This conveniently coincides with the precondition of a mechanical structure of space.

General Relativity implies that space is curved by the existence of energy (mass energy or electromagnetic energy and etc.). General Relativity is based on Riemannian geometry. If we admit this space curvature, space is assumed as an elastic body. According to continuum mechanics, the elastic body has the property of the motion of deformation such as expansion, contraction, elongation, torsion and bending.

Curvature is purely a mathematical quantity. In order to realize this mathematical curvature as a real force, the relation between strain and stress in continuum mechanics becomes important.

When we make a comparison between the space on the Earth and outer space, although there seems to be no difference, obviously a different phenomenon occurs. Simply put, an object moves radially inward, that is, drops straight down on the Earth, but in the outer space, the object floats and does not move.

The difference between the two phenomena can be explained by whether space is curved or not, that is, whether 20 independent components of a Riemann curvature tensor are zero or not. In essence, the existence of spatial curvature and curved extent region determine whether the object drops straight down or not. Although the spatial curvature at the surface of the Earth is very small value, i.e., $1.71 \times 10^{-23} (1/m^2)$, it is enough value to produce 1G (9.8 m/s^2) acceleration. Conversely, the spatial curvature in the universe is zero, therefore any acceleration is not produced. Accordingly, if the spatial curvature of a localized area containing object is controlled to curvature of $1.71 \times 10^{-23} (1/m^2)$ with a sufficiently large curved space area, the object moves and receives 1G acceleration in the universe. Of course, we are required to control both the magnitude of the curvature and the size of the curved space area.

The following Chapter 2 explains the space curvature and induced gravitational acceleration and Chapter 3 briefly introduces the concept of gravitational mechanism [8-13].

CHAPTER 2. Spatial Curvature and Derived Gravitational Acceleration

2.1. Overview of the Linear Approximation of Weak Static Gravitational Fields

The acceleration α and major curvature R^{00} are given by

$$R^{00} = \frac{1}{2} g^{ij} h_{00,ij}, \quad \alpha = c^2 \Gamma^{i}{}_{00} = \frac{1}{2} c^2 h_{00,i}, \tag{2.1}$$

respectively from the weak field approximation of the gravitational field equation.

Here, h_{00} is deviation between metric tensor g_{00} of curved space and Minkowski metric tensor η_{00} of flat space, that is,

$$g_{00} = \eta_{00} + h_{00} = -1 + h_{00} \ . \tag{2.2}$$

The notation of the symbol is as follows:

$$h_{00,ij} = \partial_i \partial_j h_{00} = \frac{\partial h_{00}}{\partial x^i \partial x^j} \ . \tag{2.3}$$

As is well known, the partial derivative $u_{i,j} = \partial_j u_i = \dfrac{\partial u_i}{\partial x^j}$ is not tensor equation. The covariant derivative $u_{i;j} = u_{i,j} - u_k \Gamma^k_{ij}$ is tensor equation and can be carried over into all coordinate systems.

If the gravitational field is time-invariant, or static, and the gravitational field is not very strong, Ricci tensor $R_{\mu\nu}$ is given by:

$$R_{\mu\nu} = \frac{1}{2}\left(\Box h_{\mu\nu} + h_{,\mu\nu} - h_{\mu\rho,\rho\nu} - h^{\rho}_{\nu,\mu\rho}\right) = \frac{1}{2}\left(\Box h_{\mu\nu} + \partial_\mu \partial_\nu h - \partial_\rho \partial_\nu h_{\mu\rho} - \partial_\mu \partial_\rho h^{\rho}_{\nu}\right) \ ,$$

$$\text{where} \quad \Box = \eta^{\mu\nu}\partial_\mu\partial_\nu = \nabla^2 - (\partial_0)^2 \quad . \tag{2.4}$$

Since all are static $h_{\mu\nu,0} = 0$ ($\partial_0 h_{\mu\nu} = 0$) and now setting $\mu = \nu = 0$, this component R_{00} is obtained

$$R_{00} = \frac{1}{2}\left(\nabla^2 h_{00} - (h_{00,0})^2 + h_{,00} - h_{0\rho,\rho 0} - h^\rho_{0,0\rho}\right) = \frac{1}{2}\nabla^2 h_{00} \quad . \tag{2.5}$$

On the other hand,

$$g_{00} = \eta_{00} + h_{00} = -1 + h_{00} = -1 - \frac{2}{c^2}\phi \quad . \tag{2.6}$$

As is well known, potential ϕ is

$$\phi = -\frac{GM}{R} \quad , \tag{2.7}$$

where M is the Earth mass, R is the Earth radius, G is the Gravity constant. We get,

$$h_{00} = -\frac{2}{c^2}\phi = -\frac{2}{c^2} \times -\frac{GM}{R} = \frac{2GM}{c^2 R} \quad . \tag{2.8}$$

Then,

$$R_{00} = \frac{1}{2}\nabla^2 h_{00} = \frac{1}{2}\nabla^2\left(-\frac{2\phi}{c^2}\right) = -\frac{1}{c^2}\nabla^2\phi \quad . \tag{2.9}$$

Curvature R_{00} can be described by the following approximation:

$$R_{00} = \frac{1}{2}\nabla^2 h_{00} = \frac{1}{2}\left(\frac{\partial^2}{\partial x^2}h_{00} + \frac{\partial^2}{\partial y^2}h_{00} + \frac{\partial^2}{\partial z^2}h_{00}\right) = \frac{1}{2}\frac{d^2 h_{00}}{dx^2} \approx \frac{1}{2}h_{00}\Big/R^2 , \tag{2.10}$$

where x is toward the Earth center.

A similar result is obtained from Eq.(2.1) as:

$$R^{00} = 1/2 \cdot g^{ij} h_{00,ij} = 1/2 \cdot g^{33}\partial^2 h_{00}/\partial x^3 \partial x^3 = 1/2 \cdot g^{33}\partial^2 h_{00}/\partial x \partial x = 1/2 \cdot \partial^2 h_{00}/\partial r^2 \approx 1/2 \cdot h_{00}/R^2$$

$$\tag{2.11}$$

The approximate expression for gravitational acceleration is:

$$\alpha = \frac{1}{2}c^2 h_{00,i} = \frac{1}{2}c^2 h_{00,x} = \frac{1}{2}c^2 \frac{dh_{00}}{dx} \approx \frac{1}{2}c^2 h_{00}\Big/R \quad . \qquad (2.12)$$

On the other hand, as will be described in detail later, the gravitational acceleration is also given by the following equation:

$$\alpha = \sqrt{-g_{00}}\,c^2 \int_a^b R^{00}(r)dr \ . \qquad (2.13)$$

Considering $g_{00} = -1$, substituting Eq. (2.10) or Eq. (2.11) into Eq. (2.13), we get

$$\alpha = c^2 \int_R^\infty \frac{1}{2}\frac{h_{00}}{r^2}dr = \frac{c^2}{2}\frac{h_{00}}{R} \qquad (2.14)$$

Eq. (2.14) matches Eq. (2.12), and the equation of gravitational acceleration expressed by Eq. (2.13) gives the mechanism of gravitation. This physical concept becomes clear in the next section.

Further, major curvature of Ricci tensor ($\mu = \nu = 0$) is calculated as follows:

$$R^{00} = g^{00}g^{00}R_{00} = -1 \times -1 \times R_{00} = R_{00} \ . \qquad (2.15)$$

Here for convenience, raise the index and use it in the notation of R^{00} instead of R_{00}.

$$R^{00} = \frac{1}{2}g^{ij}h_{00,ij} = \frac{1}{c^2}g^{ij}\left(\frac{1}{2}c^2 h_{00,i}\right)_{,j} = \frac{1}{c^2}g^{ij}\alpha_{i,j} = \frac{1}{c^2}\alpha^j_{,j}, \qquad (2.16)$$

$$\text{where } h_{00,ij} = \frac{\partial h_{00}}{\partial x^i \partial x^j} \ .$$

2.2. Gravitational Acceleration on the Earth's Surface and Its Generation Mechanism

The accumulation of surface forces in a curved area of space from the Earth's surface R to the point at infinity (∞) gives gravitational acceleration on the Earth's surface.

$$\alpha = c^2 \int_R^\infty R^{00}(r)\,dr = c^2 \int_R^\infty \frac{1}{2} h_{00} \frac{1}{r^2}\,dr = -\frac{1}{2} c^2 h_{00} \left[\frac{1}{r}\right]_R^\infty = -\frac{1}{2} c^2 h_{00}\left(0 - \frac{1}{R}\right) = \frac{1}{2}\frac{c^2 h_{00}}{R} \quad (2.17)$$

From Eq. (2.8), the deviation h_{00} of the metric tensor g_{00} from the flat space

($\eta_{00}=-1$) on the Earth's surface becomes:

$$h_{00} = \frac{2GM}{c^2 R} \qquad . \qquad\qquad (2.18)$$

The curvature of the space is (see Eq.2.10):

$$R_{00} = \left(\frac{1}{2} h_{00}\right)/R^2 \qquad . \qquad\qquad (2.19)$$

The gravitational acceleration is (see Eq.2.12):

$$\alpha = \left(\frac{1}{2} c^2 h_{00}\right)/R \qquad . \qquad\qquad (2.20)$$

Substitute the values of the Earth radius $R=6.378\times10^3$km, $GM=3.986\times10^5$km^3/s^2, c=3$\times10^5$km,

we get the following values respectively:

$$h_{00} = \frac{2GM}{c^2 R} = \frac{2\times3.986\times10^5}{(3\times10^5)^2 \times 6.378\times10^3} = \frac{2\times3.986\times10^5}{9\times6.378\times10^{13}} = 1.389\times10^{-9}$$

$$R_{00} = \left(\frac{1}{2} h_{00}\right)\times\frac{1}{R^2} = \frac{1}{2}\times1.389\times10^{-9}\times\frac{1}{(6.378\times10^3)^2} = 1.71\times10^{-2}\times10^{-9}\times10^{-6}$$

$$= 1.71\times10^{-17}(1/km)^2 = 1.71\times10^{-23}(1/m^2)$$

$$\alpha = \left(\frac{1}{2} c^2 h_{00}\right)\times\frac{1}{R} = \frac{1}{2}\times(3\times10^5)^2\times1.389\times10^{-9}\times\frac{1}{6.378\times10^3}$$

$$= \frac{1}{2}\times9\times1.389\times\frac{1}{6.378}\times10^{10}\times10^{-9}\times10^{-3} = 0.98\times10^{-2}\,km/s^2 = 9.8m/s^2$$

In this way, the following approximate values can be obtained:

The amount of displacement of the space on the Earth's surface:

$$h_{00} = 1.389\times10^{-9} .$$

Curvature of space on the Earth's surface: $R_{00} = 1.71\times10^{-23}/m^2$.

Gravitational acceleration on the Earth's surface: $\alpha = 9.8m/s^2$.

Next, a description will be given using the drawings.

Fig. 2.1 shows a concentric curved space area around the Earth. Distance of radius R from the center of the Earth is the surface of the Earth. At a distance from the Earth to an infinite point, the space becomes flat space without being affected by the gravitation of the Earth. The point at infinity is indicated by a symbol ∞ and a dotted line.

The accumulation of surface forces in a curved area of space from the Earth's surface R to the point at infinity (∞) gives gravitational acceleration on the Earth's surface, i.e., $\alpha_R = 9.8m/s^2$.

$$\alpha = c^2 \int_R^\infty R^{00}(r)dr = c^2 \int_R^\infty \frac{1}{2} h_{00} \frac{1}{r^2} dr = -\frac{1}{2} c^2 h_{00} \left[\frac{1}{r}\right]_R^\infty = -\frac{1}{2} c^2 h_{00} (0 - \frac{1}{R}) = \frac{1}{2} \frac{c^2 h_{00}}{R} \quad (2.21)$$

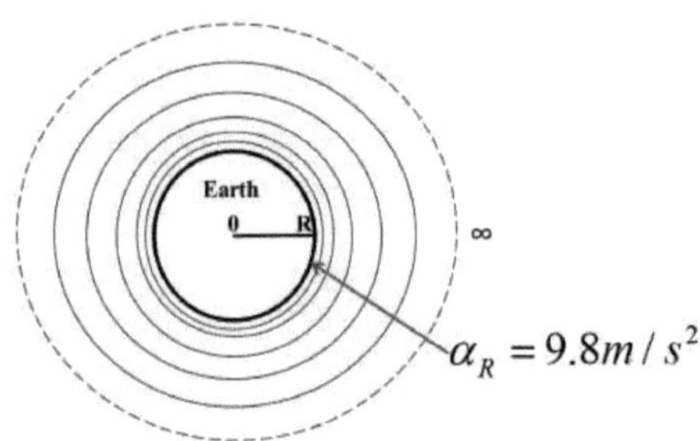

Fig. 2.1. **Mechanism of gravitational acceleration generation on the Earth's surface**

As well, Fig. 2.2 shows a concentric curved space area around the Earth. Distance of radius R from the center of the Earth is the surface of the

Earth. Here consider the gravitational acceleration at a height **h** away from the Earth's surface.

At a distance from the Earth to an infinite point, the space becomes flat space without being affected by the gravitation of the Earth. The point at infinity is indicated by a symbol ∞ and a dotted line.

The accumulation of surface forces in a curved area of space from the Earth's surface **R+h** to the point at infinity (∞) gives gravitational acceleration at the Earth's height **h**, i.e., $\alpha_{R+h} < \alpha_R = 9.8m/s^2$.

$$\alpha = c^2 \int_{R+h}^{\infty} R^{00}(r)dr = c^2 \int_{R+h}^{\infty} \frac{1}{2} h_{00} \frac{1}{r^2} dr = -\frac{1}{2} c^2 h_{00} \left[\frac{1}{r}\right]_{R+h}^{\infty} = -\frac{1}{2} c^2 h_{00} (0 - \frac{1}{R+h}) = \frac{1}{2} \frac{c^2 h_{00}}{(R+h)}$$

$$(2.22)$$

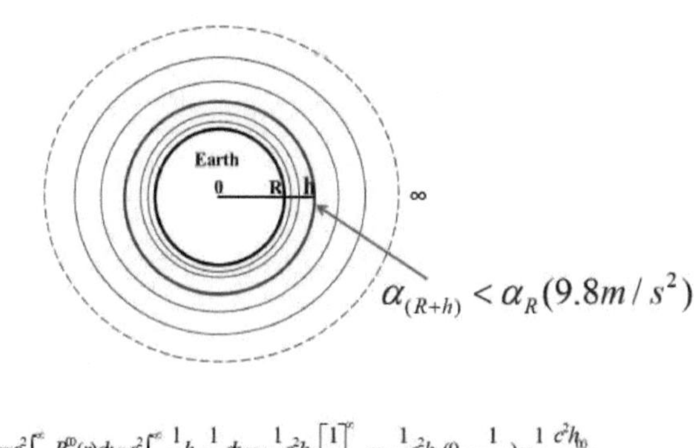

$$\alpha_{(R+h)} < \alpha_R (9.8m/s^2)$$

Fig. 2.2. Mechanism of gravitational acceleration generation on the Earth's surface height "h"

Next, consider the universal gravitational force of a well-known apple falling to the Earth.

Although the attraction between the Earth and the apple by universal gravitation can be explained by a mathematical formula, $F = G\dfrac{Mm}{r^2}$, there is no explanation of the mechanism of the attraction, that is, the principle of operation. The mechanism can be understood by interpreting that the Earth and the apple are pushed toward each other from behind the curved space area around the Earth and the curved space area around the apple.

A phenomenon is that an apple is not pulled and falls by the Earth, but the apple is pushed toward the Earth under the pressure of the vast curved space area of the Earth. Fig. 2.3 shows the mechanism.

Apple is pushed from the vast curved space area of the Earth and goes straight to the Earth. On the other hand, the Earth is also pushed from the narrow curved space area of the apple and goes straight to the apple. Since the mass of an apple is smaller than that of the Earth, the range of the curved space is small and the acceleration with respect to the Earth is almost zero.

In effect, it looks like an apple is pulled by the Earth and falls.

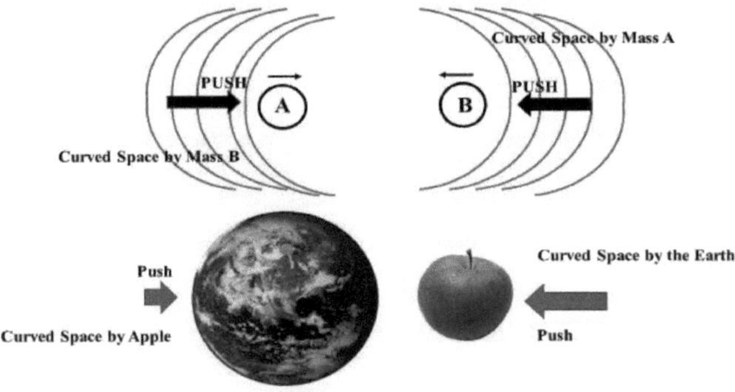

Fig. 2.3. Apple and the Earth are pushed out of a curved space and collide

Concerning the detail theory, refer to the appendix A and B for an explanation of why curved space and its curved space regions generate acceleration fields. Further, mechanical concept of space-time is introduced in the appendix C.

Appendix A: Generation of Surface Force Induced by Spatial Curvature

Appendix B: Acceleration induced by Spatial Curvature

Appendix C: Mechanical Concept of Space-time

2.3. Brief Concept of Pressure Field for Gravitational Acceleration

However, here is a brief explanation of the concept using Appendix A. Fig. 2.4 shows the fundamental principle of curved space.

On the supposition that space is an infinite continuum, continuum mechanics can be applied to the so-called "vacuum" of space. This means that space can be considered as a kind of transparent field with elastic properties.

If space curves, then an inward normal stress "$-P$" is generated. This normal stress, i.e., surface force serves as a sort of pressure field.

$$-P = N \cdot (2R^{00})^{1/2} = N \cdot (1/R_1 + 1/R_2), \qquad (2.23)$$

where N is the line stress, R_1, R_2 are the radius of principal curvature of curved surface, and R^{00} is the major component of spatial curvature.

A large number of curved thin layers form the unidirectional surface force, i.e. acceleration field. Accordingly, the spatial curvature R^{00} produces the acceleration field α.

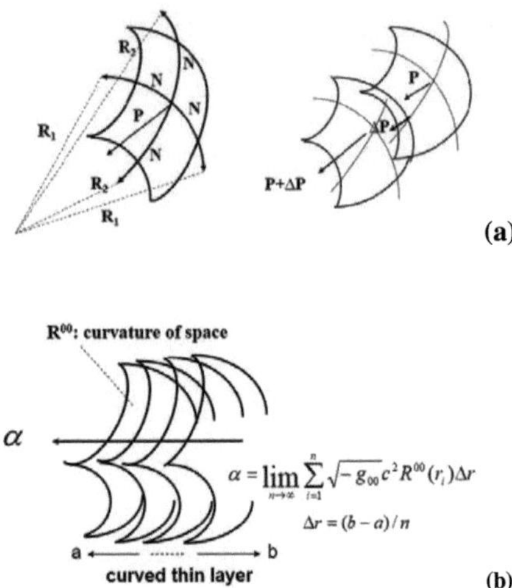

Fig. 2.4. Curvature of Space. (a) curvature of space plays a significant role. If space curves, then inward stress (surface force) "P" is generated $\Rightarrow$ A sort of pressure field; (b) a large number of curved thin layers form the unidirectional surface force, i.e. acceleration field α.

The fundamental three-dimensional space structure is determined by quadratic surface structure. Therefore, a Gaussian curvature K in two-dimensional Riemann space is significant. The relationship between K and the major component of spatial curvature R^{00} is given by:

$$K = \frac{R_{1212}}{(g_{11}g_{22} - g_{12}^{2})} = \frac{1}{2} \cdot R^{00} \quad , \qquad (2.24)$$

where R_{1212} is non-zero component of Riemann curvature tensor.

It is now understood that the membrane force on the curved surface and each principal curvature generates the normal stress "$-P$" with its direction normal to the curved surface as a surface force. The normal stress "–P" acts towards the inside of the surface as shown in Fig. 2.4 (a).

A thin-layer of curved surface will take into consideration within a spherical space having a radius of R and the principal radii of curvature that are equal to the radius ($R_1=R_2=R$). Since the membrane force N (serving as the line stress) can be assumed to have a constant value, Eq. (2.23) indicates that the curvature R^{00} generates the inward normal stress P of the curved surface. The inwardly directed normal stress serves as a pressure field.

When the curved surfaces are included in a great number, some type of unidirectional pressure field is formed. A region of curved space is made of a large number of curved surfaces and they form the field as a unidirectional surface force (i.e. normal stress). Since the field of the surface force is the field of a kind of force, the force accelerates matter in the field, i.e., we can regard the field of the surface force as the acceleration field. A large number of curved thin layers form the unidirectional acceleration field (Fig. 2.4 (b)). Accordingly, the spatial curvature R^{00} produces the acceleration field α. Therefore, the curvature of space plays a significant role to generate pressure field.

For example, consider a soap bubble.
The pressure "P" due to the membrane force on the surface of a soap bubble of radius R is directed inward. The membrane force on the surface of the soap bubble corresponds to N in the Fig. 2.4 (a).

$$-P = N \cdot (1/R_1 + 1/R_2) = N \cdot (1/R + 1/R) = 2N/R. \tag{2.25}$$

This pressure "P" keeps the soap bubbles from breaking due to the expansion force of the internal air.

Refer to the **Appendix A: Generation of Surface Force Induced by Spatial Curvature** in detail.

CHAPTER 3. Consideration of Gravitational Mechanism

Let us consider about gravitation. Why does apple fall in the Earth? A well-known answer is that there exists universal gravitation between Earth and apple. Apple is because it's pulled by a law of universal gravitation $F = G\dfrac{Mm}{r^2}$ to the Earth. Here, M is the mass of the Earth, m is the mass of apple, G is the gravitational constant, r is the distance between the Earth and apple, F is the gravitational force. From a phenomenological standpoint, it is a sufficient explanation.

However, what is the mechanism? According to General Relativity, it is said that apple moves geodesic line formed by curved space near the Earth. This is seen as lacking in sufficient explanation. The following explanation may allow someone to understand the mechanism of gravitation.

If we were to visualize the curvature of space around the Earth (M), we would describe it as having an aggregation of curved surface. A great number of thin curved surfaces are arranged in a spherical concentric pattern. This curvature would gradually become smaller as we moved away from the Earth in what we could imagine as layers of an onion. The surrounding space becomes a flat space of curvature of 0 at an imagined immense distance from the Earth (Fig. 3.1).

In the following thought experiment, an apple of mass m positioned at a distance r apart from the Earth would receive a pressure of the field formed by an accumulation of the normal stress (Fig. 3.1). As was described earlier, with reference to Fig. 2.4, the membrane force on the curved surface and each principal curvature generates the normal stress"$-P$" with its direction normal to the curved surface as a surface force. The normal stress "$-P$" acts towards the inside of the surface as shown in Fig. 2.4 (a).

A thin-layer of curved surface will take into consideration within a spherical space having a radius of R and the principal radii of curvature that are equal to the radius ($R_1=R_2=R$).Since the membrane force N (serving as the line stress) can be assumed to have a constant value, the inwardly directed normal stress serves as a pressure field. When the curved surfaces are included in a great number, some type of unidirectional pressure field is formed. That is, a sort of graduated pressure field is generated by the curved range from an arbitrary point "a" in curved space to a point "b" (the point at which space is absent of curvature, i.e., flat space of curvature 0) (Fig. 3.1). Then apple moves directly towards the center of the Earth, that is, the apple falls. Falling acceleration of apple in curved space is proportional to both the value of spatial curvature and the size of curved space.

If the Earth (M) were to disappear instantly, the curved surface of space close to the Earth would return to the flat surface (see Fig. 3.2). Because an external action causing curvature (i.e., mass energy) disappears. The change from a curved surface to a flat surface would advance the position r of the apple at the speed of light (i.e., the strain rate of space-time).

The propagation velocity of the change from flat space to curved space and the propagation velocity of change from curved space to flat space are both the same, i.e. the velocity of light.

However, in our thought experiment, the apple would still receive pressure from the surrounding field by the accumulation of the normal stress.

Because, since there still exists the curved region behind the apple from "a" to "b" (the remote flat space), the apple continues falling. The pressure continues to push the apple to the center of the Earth (Fig. 3.2).

However, as soon as the change from a curved surface to a flat surface passes through the point of the apple (i.e., "a" point), the pressure at point "a" disappears and the apple would only float without falling (or keep falling due to inertia) (Fig. 3.3).

The above discussion provides a basis to consider the following thought experiment. Even if the Sun instantly disappeared, the Earth would still continue to revolve around the Sun until 8 minutes 32 seconds, or the time at which it takes light to advance between the Sun and Earth. However, as soon as the change from curved surface to flat surface passes through the point of the Earth, or at 8 minutes 32 seconds after the event, the pressure pushing the Earth would disappear, and the Earth would fly away in a direction tangential to the revolution orbit.

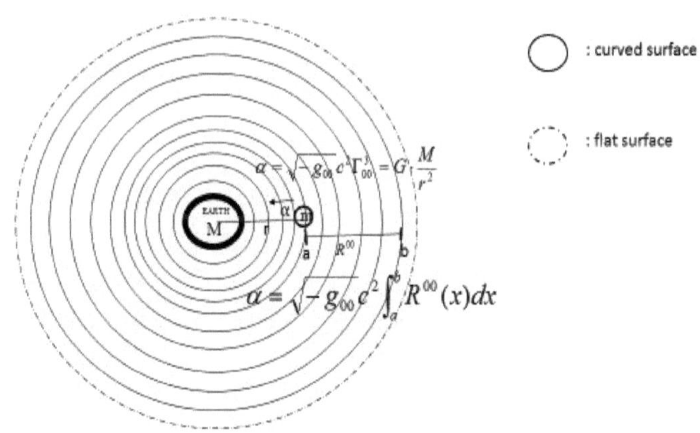

Fig. 3.1. Apple falls receiving a pressure of the field

27

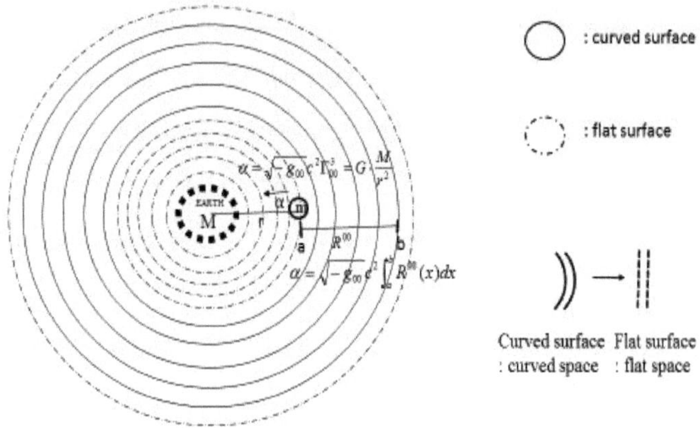

Fig. 3.2. Apple still continues falling receiving a pressure of the field

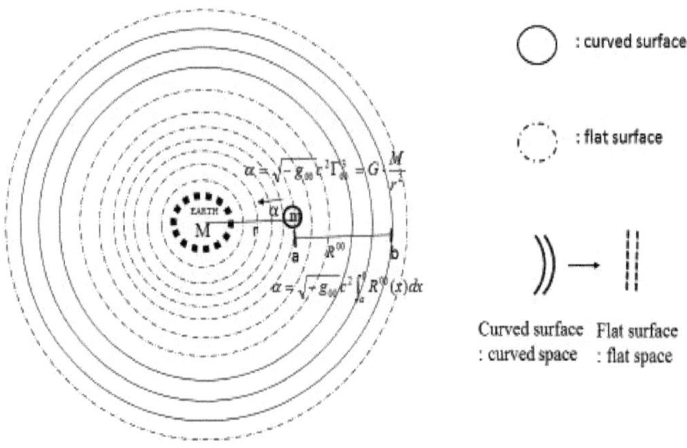

Fig. 3.3. Apple only floats without falling due to the lack of pressure of the field

In view of this, gravitation may be considered as a pressure generated in a region of curved space.

Conclusion

Assuming that space is an infinite continuum, a mechanical concept of space became identified. Space can be considered as a kind of transparent elastic field. The pressure field derived from the geometrical structure of space is newly obtained by applying both continuum mechanics and General Relativity to space. As a result, a fundamental concept of space-time is described that focuses on theoretically innate properties of space including strain and curvature.

The mass on the Earth will not be pulled by the Earth and fall, but will be pushed and fall in the direction of the Earth due to the pressure of the field in the curved space area around the Earth. Although the spatial curvature at the surface of the Earth is very small value, i.e., $1.71 \times 10^{-23} (1/m^2)$, it is enough value to produce 1G (9.8 m/s^2) acceleration.

Gravity (Gravitation) can be explained as a pressure field induced by the curvature of space.

This concept can be also applied to the gravity effect of stars, stars in the galactic universe, planets, etc.

As is well known, there are four kinds of forces in the universe: strong force, electromagnetic force, weak force, and gravity. Since the space-time is created at the beginning after the birth of the universe, gravity will be generated compared to other forces first.

Except for gravity, quantum sources have been discovered, and those forces are described in quantum theory. The underlying quantum graviton for gravity is not found, and gravity has not been described successfully in quantum theory.

Some basic concepts of quantum theory and General Relativity are incompatible. In quantum theory, one time must be set in all flat space-time to define a state. Meanwhile, in the General Relativity, time is not defined globally over the entire space-time and can be introduced only locally, which makes it difficult to connect with quantum theory. It is not possible to renormalize even if it is locally limited, and it is still unsolved, including infinite divergence.

Field quantization is performed on wave functions and gauge fields that describe elementary particles, not on space-time. In quantum field theory, space-time remains continuous. Is it possible to quantize a continuum space? In other words, is it possible to quantize gravity, which is considered to be the pressure of a space field? This book has introduced one possibility.

REFERENCES

[1] Minami, Y. 1988. "Space Strain Propulsion System." In *Proceedings of 16th International Symposium on Space Technology and Science*, 125-36.

[2] Minami, Y. 1994. "Possibility of Space Drive Propulsion." In *Proceedings of 45th Congress of the International Astronautical Federation*, 658.

[3] Minami, Y. 1997. "Spacefaring to the Farthest Shores — Theory and Technology of a Space Drive Propulsion System." *Journal of the British Interplanetary Society* 50: 263-76.

[4] Minami, Y. 2003. "An Introduction to Concepts of Field Propulsion." *Journal of the British InterplanetarySociety*56: 350-9.

[5] Minami, Y. 2013. "Space Drive Propulsion Principle from the Aspect of Cosmology." *Journal of Earth Science and Engineering* 3: 379-92.

[6] Minami, Y. 2015. "Space propulsion physics toward galaxy exploration." J Aeronaut Aerospace Eng 4: 2.

[7] Minami, Y., *STAR FLIGHT THEORY: By the Physics of Field Propulsion*, published in July. 15, 2019, (LAMBERT Academic Publishing);
https://www.morebooks.shop/store/gb/book/star-flight-theory-:-by-the-physics-of-field-propulsion/isbn/978-620-0-23433-9

[8] Flügge, W. 1972. *Tensor Analysis and Continuun Mechanics*. New York: Springer-Verlag.

[9] Fung,Y.C. 2001. *Classical and Computational Solid Mechanics*. New Jersey: World Scientific Publishing Co. Pre. Ltd..

[10] Borg, S. F. 1963. *Matrix-Tensor Methods in Continuum Mechanics*. New York: D. Van Nostrand Company.

[11] Williams, C. (Editor); Minami, Y. (Chap.3); et al. *Advances in General Relativity Research*, Nova Science Publishers, 2015.

[12] Minami, Y. "Continuum Mechanics of Space Seen from the Aspect of General Relativity － An Interpretation of the Gravity Mechanism", *Journal of Earth Science and Engineering* 5, 2015: 188-202.

[13] Minami, Y. "Gravitational Effects Generated by the Curvature of Space on the Earth's Surface", *Journal of Scientific and Engineering Research*, 2020, 7(3):1-15.

APPENDICES

Appendix A: Generation of Surface Force Induced by Spatial Curvature

On the supposition that space is an infinite continuum, continuum mechanics can be applied to the so-called "vacuum" of space. This means that space can be considered as a kind of transparent field with elastic properties.

If space curves, then an inward normal stress " $-P$ " is generated. This normal stress, i.e. surface force serves as a sort of pressure field.

$$-P = N \cdot (2R^{00})^{1/2} = N \cdot (1/R_1 + 1/R_2) \tag{A1}$$

where N is the line stress, R_1, R_2 are the radius of principal curvature of curved surface, and R^{00} is the major component of spatial curvature.

A large number of curved thin layers form the unidirectional surface force, i.e. acceleration field. Accordingly, the spatial curvature R^{00} produces the acceleration field α.

The fundamental three-dimensional space structure is determined by quadratic surface structure. Therefore, a Gaussian curvature K in two-dimensional Riemann space is significant. The relationship between K and the major component of spatial curvature R^{00} is given by:

$$K = \frac{R_{1212}}{(g_{11}g_{22} - g_{12}{}^2)} = \frac{1}{2} \cdot R^{00}, \tag{A2}$$

where R_{1212} is non-zero component of Riemann curvature tensor.

<Reference matters for Eq.(A2)>

For a two-dimensional surface, from the Bianchi identity, the Riemann curvature tensor is given by $R_{\nu\mu\lambda\kappa} = K(g_{\nu\lambda}g_{\mu\kappa} - g_{\nu\kappa}g_{\mu\lambda})$, that is,

$$R_{1212} = K(g_{11}g_{22} - g_{12}{}^2).$$

And, for a spherical surface of radius r, its Gaussian curvature K is 1 / r^2.

The scalar curvature R and the Gaussian curvature K on the quadratic surface are as follows:

$$R = R_i{}^i = g^{ij}R_{ij} = g^{11}R_{11} + g^{22}R_{22} = \frac{1}{r^2}(-1) + \frac{1}{r^2\sin^2\theta}(-\sin^2\theta) = -\frac{2}{r^2} = -2K .$$

(Using $R_{\mu\nu} = R^{\sigma}{}_{\mu\sigma\nu} = g^{\rho\sigma}R_{\rho\mu\sigma\nu}$, calculate the rich tensor, the following are

obtained: $R_{11} = -1,\quad R_{22} = -\sin^2\theta,\quad R_{12} = R_{21} = 0$.)

The scalar curvature R $(1/m^2)$ on a four-dimensional surface is given by

$$R = R_i{}^i = g_{ij}R^{ij} = g_{00}R^{00} + g_{11}R^{11} + g_{22}R^{22} + g_{33}R^{33} \approx g_{00}R^{00} = -R^{00}\ (g_{00} \approx -1: weak\ field)$$

Thus, $K = \frac{1}{2} \cdot R^{00}$ is obtained.

It is now understood that the membrane force on the curved surface and each principal curvature generates the normal stress "$-P$" with its direction normal to the curved surface as a surface force. The normal stress "$-P$" acts towards the inside of the surface as shown in Fig.A1 (a).

A thin-layer of curved surface will take into consideration within a spherical space having a radius of R and the principal radii of curvature that are equal to the radius ($R_1 = R_2 = R$). Since the membrane force N (serving as the line stress) can be assumed to have a constant value, Eq. (A1) indicates that the curvature R^{00} generates the inward normal stress P of the curved surface. The inwardly directed normal stress serves as a pressure field.

When the curved surfaces are included in a great number, some type of unidirectional pressure field is formed. A region of curved space is made of a large number of curved surfaces and they form the field as a unidirectional surface force (i.e. normal stress). Since the field of the surface force is the field of a kind of force, the force accelerates matter in the field, i.e., we can regard the field of the surface force as the acceleration field. A large number of curved thin layers form the unidirectional acceleration field (Fig. A1 (b)). Accordingly, the spatial curvature R^{00} produces the acceleration field α. Therefore, the curvature of space plays a significant role to generate pressure field.

Applying membrane theory, the following equilibrium conditions are obtained in quadratic surface, given by:

$$N^{\alpha\beta} b_{\alpha\beta} + P = 0 \quad , \tag{A3}$$

where $N^{\alpha\beta}$ is a membrane force, i.e. line stress of curved space, $b_{\alpha\beta}$ is second fundamental metric of curved surface, and P is the normal stress on curved surface [8].

The second fundamental metric of curved space $b_{\alpha\beta}$ and principal curvature $K_{(i)}$ has the following relationship using the metric tensor $g_{\alpha\beta}$,

$$b_{\alpha\beta} = K_{(i)} g_{\alpha\beta} \quad . \tag{A4}$$

Therefore, we get:

$$N^{\alpha\beta} b_{\alpha\beta} = N^{\alpha\beta} K_{(i)} g_{\alpha\beta} = g_{\alpha\beta} N^{\alpha\beta} K_{(i)} = N_\alpha{}^\alpha K_{(i)} = N \cdot K_{(i)} . \tag{A5}$$

From Eq. (A3) and Eq. (A5), we get:

$$N_\alpha{}^\alpha K_{(i)} = -P . \tag{A6}$$

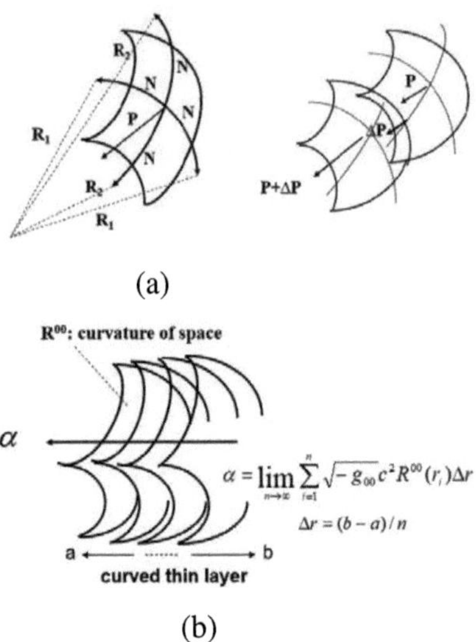

(a)

(b)

Fig. A1. Curvature of Space: **(a)** curvature of space plays a significant role. If space curves, then inward stress (surface force) "P" is generated $\Rightarrow$ A sort of pressure field; **(b)** a large number of curved thin layers form the unidirectional surface force, i.e. acceleration field α.

As for the quadratic surface, the indices α and i take two different values, i.e. 1 and 2, therefore Eq. (A6) becomes:

$$N_1{}^1 K_{(1)} + N_2{}^2 K_{(2)} = -P \quad , \tag{A7}$$

where $K_{(1)}$ and $K_{(2)}$ are principal curvature of curved surface and are inverse number of radius of principal curvature (i.e. $1/R_1$ and $1/R_2$).

The Gaussian curvature K is represented as:

$$K = K_{(1)} \cdot K_{(2)} = (1/R_1) \cdot (1/R_2) \quad . \tag{A8}$$

Accordingly, suppose $N_1^{\;1} = N_2^{\;2} = N$, we get:

$$N \cdot (1/R_1 + 1/R_2) = -P \quad . \tag{A9}$$

It is now understood that the membrane force on the curved surface and each principal curvature generate the normal stress "$-P$" with its direction normal to the curved surface as a surface force. The normal stress "$-P$" is towards the inside of surface as showing in Fig. A1(a).

A thin-layer of curved surface will be taken into consideration within a spherical space having a radius of R and the principal radii of curvature which are equal to the radius ($R_1=R_2=R$). From Eqs. (A2) and (A8), we then get:

$$K = \frac{1}{R_1} \cdot \frac{1}{R_2} = \frac{1}{R^2} = \frac{R^{00}}{2} \quad . \tag{A10}$$

Considering $N \cdot (2/R) = -P$ of Eq.(A9), and substituting Eq. (A10) into Eq. (A9), the following equation is obtained:

$$-P = N \cdot \sqrt{2R^{00}} \quad . \tag{A11}$$

Since the membrane force N (serving as the line stress) can be assumed to have a constant value, Eq. (A11) indicates that the curvature R^{00} generates the inward normal stress P of the curved surface. The inwardly directed normal stress serves as a kind of pressure field. Accordingly, the cumulated curved region of curvature R^{00} produces the acceleration field α.

Here, we give an account of curvature R^{00} in advance. The solution of metric tensor $g^{\mu\nu}$ is found by gravitational field equation as the following:

$$R^{\mu\nu} - \frac{1}{2} \cdot g^{\mu\nu} R = -\frac{8\pi G}{c^4} \cdot T^{\mu\nu} \quad , \tag{A12}$$

where $R^{\mu\nu}$ is the Ricci tensor, R is the scalar curvature, G is the gravitational constant, c is the velocity of light, $T^{\mu\nu}$ is the energy momentum tensor.

Furthermore, we have the following relation for scalar curvature R :

$$R = R^{\alpha}{}_{\alpha} = g^{\alpha\beta} R_{\alpha\beta}, \quad R^{\mu\nu} = g^{\mu\alpha} g^{\nu\beta} R_{\alpha\beta}, \quad R_{\alpha\beta} = R^{j}{}_{\alpha j\beta} = g^{ij} R_{i\alpha j\beta} \quad . \tag{A13}$$

Ricci tensor $R_{\mu\nu}$ is also represented by:

$$R_{\mu\nu} = \Gamma^{\alpha}_{\mu\alpha,\nu} - \Gamma^{\alpha}_{\mu\nu,\alpha} - \Gamma^{\alpha}_{\mu\nu}\Gamma^{\beta}_{\alpha\beta} + \Gamma^{\alpha}_{\mu\beta}\Gamma^{\beta}_{\nu\alpha} \quad (= R_{\nu\mu}) , \tag{A14}$$

where $\Gamma^{i}{}_{jk}$ is Riemannian connection coefficient.

If the curvature of space is very small, the term of higher order than the second can be neglected, and Ricci tensor becomes:

$$R_{\mu\nu} = \Gamma^{\alpha}_{\mu\alpha,\nu} - \Gamma^{\alpha}_{\mu\nu,\alpha} \quad . \tag{A15}$$

The major curvature of Ricci tensor ($\mu = \nu = 0$) is calculated as follows:

$$R^{00} = g^{00} g^{00} R_{00} = -1 \times -1 \times R_{00} = R_{00} \quad . \tag{A16}$$

As previously mentioned, Riemannian geometry is a geometry that deals with a curved Riemann space, therefore a Riemann curvature tensor is the principal quantity. All components of Riemann curvature tensor are zero for flat space and non-zero for curved space. If an only non-zero component of Riemann curvature tensor exists, the space is not flat space but curved space. Therefore, the curvature of space plays a significant role.

Appendix B: Acceleration Induced by Spatial Curvature

A massive body causes the curvature of space-time around it, and a free particle responds by moving along a geodesic line in that space-time. The path of free particle is a geodesic line in space-time and is given by the following geodesic equation;

$$\frac{d^2 x^i}{d\tau^2} + \Gamma^i_{jk} \cdot \frac{dx^j}{d\tau} \cdot \frac{dx^k}{d\tau} = 0, \tag{B1}$$

where Γ^i_{jk} is Riemannian connection coefficient, τ is proper time, x^i is four-dimensional Riemann space, that is, three dimensional space (x=x^1, y=x^2, z=x^3) and one dimensional time (w=ct=x^0), c is the velocity of light. These four coordinate axes are denoted as x^i (i=0, 1, 2, 3).

Proper time is the time to be measured in a clock resting for a coordinate system. We have the following relation derived from an invariant line element ds^2 between Special Relativity (flat space) and General Relativity (curved space):

$$d\tau = \sqrt{-g_{00}}\, dx^0 = \sqrt{-g_{00}}\, cdt. \tag{B2}$$

From Eq. (B1), the acceleration of free particle is obtained by

$$\alpha^i = \frac{d^2 x^i}{d\tau^2} = -\Gamma^i_{jk} \cdot \frac{dx^j}{d\tau} \cdot \frac{dx^k}{d\tau}. \tag{B3}$$

As is well known in General Relativity, in the curved space region, the massive body "m (kg)" existing in the acceleration field is subjected to the following force F^i(N) :

$$F^i = m\Gamma^i_{jk} \cdot \frac{dx^j}{d\tau} \cdot \frac{dx^k}{d\tau} = m\sqrt{-g_{00}}\, c^2 \Gamma^i_{jk} u^j u^k = m\alpha^i, \tag{B4}$$

where u^j, u^k are the four velocity, Γ^i_{jk} is the Riemannian connection coefficient, and τ is the proper time.

From Eqs. (B3),(B4), we obtain:

$$\alpha^i = \frac{d^2x^i}{d\tau^2} = -\Gamma^i_{jk} \cdot \frac{dx^j}{d\tau} \cdot \frac{dx^k}{d\tau} = -\sqrt{-g_{00}}\, c^2 \Gamma^i_{\ jk} u^j u^k \quad . \tag{B5}$$

Eq. (B5) yields a more simple equation from the condition of linear approximation, that is, weak-field, quasi-static, and slow motion (speed v << speed of light $c : u^0 \approx 1$):

$$\alpha^i = -\sqrt{-g_{00}} \cdot c^2 \Gamma^i_{00} \quad . \tag{B6}$$

On the other hand, the major component of spatial curvature R^{00} in the weak field is given by

$$R^{00} \approx R_{00} = R^\mu_{0\mu 0} = \partial_0 \Gamma^\mu_{0\mu} - \partial_\mu \Gamma^\mu_{00} + \Gamma^\nu_{0\mu}\Gamma^\mu_{\nu 0} - \Gamma^\nu_{00}\Gamma^\mu_{\nu\mu} \quad . \tag{B7}$$

In the nearly Cartesian coordinate system, the value of $\Gamma^\mu_{\nu\rho}$ are small, so we can neglect the last two terms in Eq. (B7), and using the quasi-static condition we get

$$R^{00} = -\partial_\mu \Gamma^\mu_{00} = -\partial_i \Gamma^i_{00} \quad . \tag{B8}$$

From Eq. (B8), we get formally

$$\Gamma^i_{00} = -\int R^{00}(x^i)dx^i \quad . \tag{B9}$$

Substituting Eq. (B9) into Eq. (B6), we obtain

$$\alpha^i = \sqrt{-g_{00}}\, c^2 \int R^{00}(x^i)dx^i \quad . \tag{B10}$$

Accordingly, from the following linear approximation scheme for the gravitational field equation: (1) weak gravitational field, i.e. small curvature limit, (2) quasi-static, (3) slow-motion approximation (i.e., $v/c \ll 1$), and considering range of curved region, we get the following relation between acceleration of curved space and curvature of space:

$$\alpha^i = \sqrt{-g_{00}}\, c^2 \int_a^b R^{00}(x^i)\, dx^i \;, \tag{B11}$$

where α^i: acceleration (m/s^2), g_{00}: time component of metric tensor, a-b: range of curved space (m), x^i: components of coordinate (i=0,1,2,3), c: velocity of light, R^{00}: major component of spatial curvature (1/m^2).

Eq. (B11) indicates that the acceleration field α^i is produced in curved space. The intensity of acceleration produced in curved space is proportional to the product of spatial curvature R^{00} and the length of curved region.

Eq. (B4) yields more simple equation from above-stated linear approximation ($u^0 \approx 1$),

$$F^i = m\sqrt{-g_{00}}\, c^2 \Gamma_{00}^i u^0 u^0 = m\sqrt{-g_{00}}\, c^2 \Gamma_{00}^i = m\alpha^i = m\sqrt{-g_{00}}\, c^2 \int_a^b R^{00}(x^i)\, dx^i \;. \tag{B12}$$

Setting i=3 (i.e., direction of radius of curvature: r), we get Newton's second law:

$$F^3 = F = m\alpha = m\sqrt{-g_{00}}\, c^2 \int_a^b R^{00}(r)\, dr = m\sqrt{-g_{00}}\, c^2 \Gamma_{00}^3 \;. \tag{B13}$$

The acceleration (α) of curved space and its Riemannian connection coefficient (Γ_{00}^3) are given by:

$$\alpha = \sqrt{-g_{00}}\, c^2 \Gamma_{00}^3 \;, \qquad \Gamma_{00}^3 = \frac{-g_{00,3}}{2g_{33}} \;, \tag{B14}$$

where c: velocity of light, g_{00} and g_{33}: component of metric tensor, $g_{00,3}$: $\partial g_{00}/\partial x^3 = \partial g_{00}/\partial r$. We choose the spherical coordinates "$ct=x^0$, $r=x^3, \theta=x^1$, $\varphi=x^2$" in space-time. The acceleration α is represented by the equation both in the differential form and in the integral form. Practically, since the metric is usually given by the solution of gravitational field equation, the differential form has been found to be advantageous.

Now in general, the line element is described in:

$$ds^2 = g_{ij}dx^i dx^j = g_{00}(dx^0)^2 + g_{33}(dx^3)^2 + g_{11}(dx^1)^2 + g_{22}(dx^2)^2$$
$$= g_{00}(cdt)^2 + g_{33}(dr)^2 + g_{11}r^2(d\theta)^2 + g_{22}r^2 \sin^2\theta(d\varphi)^2 \qquad (B15)$$

We choose the spherical coordinates "$ct=x^0$, $r=x^3$, $\theta=x^1$, $\varphi=x^2$" in space-time.

Next, let us consider External Schwarzschild Solution.

The line element is obtained as follows:

$$ds^2 = -(1-\frac{r_g}{r})c^2 dt^2 + \frac{1}{1-\frac{r_g}{r}}dr^2 + r^2(d\theta^2 + \sin^2\theta d\varphi^2). \qquad (B16)$$

The metrics are given by:

$$g_{00} = -(1-r_g/r), g_{11} = g_{22} = 1, g_{33} = 1/(1-r_g/r),$$
$$and\ other\ g_{ij} = 0. \qquad (B17)$$

where r_g is the gravitational radius (i.e. $r_g = 2GM/c^2$).

Combining Eq.(B17) with Eq. (B14) yields:

$$\alpha = G\cdot\frac{M}{r^2}, (r_g\langle r) \ , \qquad (B18)$$

where G is a gravitational constant and M is a total mass.

Eq. (B18) indicates the gravitational acceleration of the well-known the Earth mass M. That is, universal gravitation $F = G\frac{Mm}{r^2}$.

Appendix C: Mechanical Concept of Space-time

Fundamental Concept of Space

Space is an infinite continuum and its structure is determined by Riemannian geometry. Space satisfies the following conditions:

a) When the infinitesimal distance regulating the distance between the two points changes by a certain physical action, the change is continuous, and the space maintains a continuum even after its change. Now, the concept of strain of continuum mechanics is very important in order to relate a spatial curvature to a practical force. Because the spatial curvature is a purely geometrical quantity. A strain field is required for the conversion of geometrical quantity to a practical force.

b) The spatial strain is defined as a localized geometrical structural change of space. It implies a change from flat space involved in zero curvature components to curved Riemann space involved in non-zero curvature components.

c) Space has the only strain-free natural state, and space always returns to the strain-free natural state, i.e., flat space, when an external physical action causing spatial strain is removed.

d) Spatial strain means some kinds of structural deformation of space, and a body filling up space is affected by the action from its spatial strain. We must distinguish space from an isolated body. An isolated body occupies an area of space by its movement. Basically, an isolated body can move in space and also can change its position.

e) In order to keep the continuity of space, the velocity of body filling up space cannot exceed the strain rate of space itself.

Since the subject of our study is a four-dimensional Riemann space as a curved space, we ascribe a great deal of importance to the curvature of space. We a priori accept that the nature of actual physical space is a four-dimensional Riemann space, that is, three dimensional space ($x=x^1$, $y=x^2$, $z=x^3$) and one dimensional time ($w=ct=x^0$), where c is the velocity of light. These four coordinate axes are denoted as x^i (i=0, 1, 2, 3).

The square of the infinitesimal distance "ds" between two infinitely proximate points x^i and x^i+dx^i is given by equation of the form:

$$ds^2 = g_{ij}dx^i dx^j, \qquad (C1)$$

where g_{ij} is a metric tensor.

The metric tensor g_{ij} determines all the geometrical properties of space and it is a function of this space coordinate. In Riemann space, the metric tensor g_{ij} determines a Riemannian connection coefficient Γ^i_{jk} , and furthermore determines the Riemann curvature tensor R^p_{ijk} *or* R_{pijk}, thus the geometry of space is determined by a metric tensor.

Riemannian geometry is a geometry which provides a tool to describe curved Riemann space, therefore a Riemann curvature tensor is the principal quantity. All the components of Riemann curvature tensor are zero for flat space and non-zero for curved space. If a non-zero component of Riemann curvature tensor exists, the space is not flat space, but curved space. In curved space, it is well known that the result of the parallel displacement of vector depends on the choice of the path. Further, the components of a vector differ from the initial value, after we displace a vector parallel along a closed curve until it returns to the starting point.

An external physical action such as the existence of mass energy or electromagnetic energy yields the structural deformation of space. In the deformed space region, the infinitesimal distance is given by:

$$ds'^2 = g'_{ij} dx^i dx^j ,\qquad\qquad (C2)$$

where g'_{ij} is the metric tensor of deformed space region, and we use the convected coordinates ($x'^i = x^i$).

As shown in Fig.C1, if the line element between the arbitrary two near points (A and B) in space region **S** (before structural deformation) is defined as $ds = g_i dx^i$, the infinitesimal distance between the two near points is given by Eq. (C1): $ds^2 = g_{ij} dx^i dx^j$.

Let us assume that a space region **S** is structurally deformed by an external physical action and transformed to space region **T**. In the deformed space region **T**, the line element between the identical two near point (A' and B') of the identical space region newly changes, differs from the length and direction, and becomes $ds' = g'_i dx^i$.

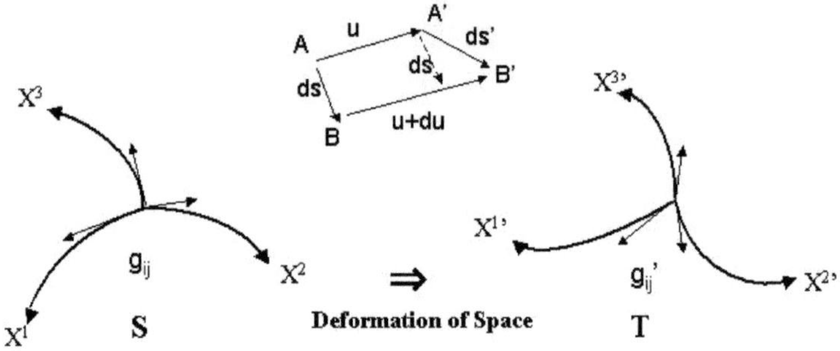

Fig. C1. Fundamental structure of Space

Therefore, the infinitesimal distance between the two near points using the convected coordinate ($x'^i = x^i$) is given by:

$$ds'^2 = g'_{ij}dx^i dx^j .\tag{C3}$$

The g'_i is the transformed base vector from the original base vector g^i and the g'_{ij} is the transformed metric tensor from the original metric tensor g_{ij}. Since the degree of deformation can be expressed as the change of distance between the two points, we get:

$$ds'^2 - ds^2 = g'_{ij}dx^i dx^j - g_{ij}dx^i dx^j = (g'_{ij} - g_{ij})dx^i dx^j = r_{ij}dx^i dx^j .\tag{C4}$$

Hence the degree of geometrical and structural deformation can be expressed by the quantity denoted change of metric tensor, i.e.

$$r_{ij} = g'_{ij} - g_{ij} .\tag{C5}$$

On the other hand, the state of deformation can be also expressed by the displacement vector "u" (see Fig.C1).

From the continuum mechanics [8], the following equations are obtained:

$$du = g^i u_{i:j} dx^j \,, \tag{C6}$$

$$ds' = ds + du = ds + g^i u_{i:j} dx^j \quad . \tag{C7}$$

We use the usual notation ":" for covariant differentiation.

As is well known, the partial derivative $u_{i,j} = \dfrac{\partial u_i}{\partial x^j}$ is not tensor equation. The covariant derivative $u_{i:j} = u_{i,j} - u_k \Gamma_{ij}^k$ is tensor equation and can be carried over into all coordinate systems.

From the usual continuum mechanics, the infinitesimal distance after deformation becomes [8]:

$$ds'^2 - ds^2 = r_{ij} dx^i dx^j = (u_{i:j} + u_{j:i} + u^k{}_{:i} u_{k:j}) dx^i dx^j \quad . \tag{C8}$$

The terms of higher order than second $u^k{}_{:i} u_{k:j}$ can be neglected if the displacement is small enough value. As the actual physical space can be dealt with the minute displacement from the trial calculation of strain, we get:

$$r_{ij} = u_{i:j} + u_{j:i} \quad . \tag{C9}$$

Whereas, according to the continuum mechanics [8], the strain tensor e_{ij} is given by:

$$e_{ij} = \frac{1}{2} \cdot r_{ij} = \frac{1}{2} \cdot (u_{i:j} + u_{j:i}) \quad . \tag{C10}$$

So, we get:

$$ds'^2 - ds^2 = (g'_{ij} - g_{ij}) dx^i dx^j = 2e_{ij} dx^i dx^j \,, \tag{C11}$$

where g'_{ij}, g_{ij} is a metric tensor, e_{ij} is a strain tensor, and $ds'^2 - ds^2$ is the square of the infinitesimal distance between two infinitely proximate points x^i and $x^i + dx^i$.

Eq.(C11) indicates that a certain geometrical structural deformation of space is shown by the concept of strain. In essence, the change of metric tensor $(g'_{ij} - g_{ij})$ due to the existence of mass energy or electromagnetic energy tensor produces the strain field e_{ij}.

Since space-time is distorted, the infinitesimal distance between two infinitely proximate points x^i and $x^i + dx^i$ is important in our understanding of the geometry of the space-time; the physical strain is generated by the difference of a geometrical metric of space-time. Namely, a certain structural deformation is described by strain tensor e_{ij}. From Eq.(C11), the strain of space is described as follows:

$$e_{ij} = 1/2 \cdot (g_{ij}' - g_{ij}) \quad . \tag{C12}$$

It is also worth noting that this result yields the principle of constancy of light velocity in Special Relativity.

Mechanics of Space

Expanding the concept of vector parallel displacement in Riemann space, the following equation has newly been obtained:

$$\omega_{\mu\nu} = R_{\mu\nu kl} dA^{kl} , \tag{C13}$$

where $\omega_{\mu\nu}$ is rotation tensor, dA^{kl} is infinitesimal areal element.

According to the nature of Riemann curvature tensor $R_{\mu\nu kl}$, $\omega_{\mu\nu}$ indicates the rotation of displacement field. Eq. (C13) indicates that a curved space produces the rotation of displacement field in the region of space. Now, the rotation tensor $\omega_{\mu\nu}$ and strain tensor e_{ij} satisfy the following differential equation in continuum mechanics:

$$\omega_{\mu\nu,j} = e_{\nu j,\mu} - e_{\mu j,\nu} \quad . \tag{C14}$$

This equation is true on condition that the order of differential can be exchanged in a flat space. To expand above equation into a curved Riemann space, the equation shall be transformed to covariant differentiation and it is possible on condition of $\Gamma^{\alpha}_{j\nu} e_{\mu\alpha} = \Gamma^{\alpha}_{j\mu} e_{\nu\alpha}$.

Thus, we obtain

$$\omega_{\mu\nu.j} = e_{\nu j:\mu} - e_{\mu j:\nu} \quad . \tag{C15}$$

Here we use the usual notation ":" for covariant differentiation.

Eq. (C15) indicates that the displacement gradient of rotation tensor corresponds to difference of the displacement gradient of strain tensor.

Here, if we multiply both sides of Eq.(C15) by fourth order tensor denoted the nature of space $E^{ij\mu\nu}$ formally, we obtain

$$E^{ij\mu\nu} \omega_{\mu\nu.j} = E^{ij\mu\nu} (R_{\mu\nu kl} dA^{kl})_{:j} = E^{ij\mu\nu} R_{\mu\nu kl:j} dA^{kl} , \tag{C16}$$

and

$$E^{ij\mu\nu} e_{\nu j:\mu} - E^{ij\mu\nu} e_{\mu j:\nu} = (E^{ij\mu\nu} e_{\nu j})_{:\mu} - (E^{ij\mu\nu} e_{\mu j})_{:\nu} = \sigma^{i\mu}{}_{:\mu} - \sigma^{i\nu}{}_{:\nu} = \Delta\sigma^{ir}{}_{:r} . \tag{C17}$$

As is well known in the continuum mechanics [8, 9, 10], the relationship between stress tensor σ_{ij} and strain tensor e_{ml} is given by

$$\sigma^{ij} = E^{ijml} e_{ml} \quad .\tag{C18}$$

Furthermore, the relationship between body force F^i and stress tensor σ_{ij} is given by

$$F^i = \sigma^{ij}{}_{:j} \, ,\tag{C19}$$

from the equilibrium conditions of continuum. That is, the elastic force F^i is given by the gradient of stress tensor σ^{ij}.

Therefore, Eq. (C17) indicates the difference of body force ΔF^i. Accordingly, from Eqs. (C16) and (C17), the change of body force $\Delta F^i (= \Delta \sigma^{ir}{}_{:r})$ becomes

$$\Delta F^i = E^{ij\mu\nu} R_{\mu\nu kl:j} dA^{kl} \quad .\tag{C20}$$

Here, we assume that $E^{ij\mu\nu}$ is constant for covariant differentiation, and A^{kl} is area element.

The stress tensor σ^{ij} is a surface force and F^i is a body force. The body force is an equivalent gravitational action because of acting all elements of space uniformly.

Eq. (C20) indicates that the gradient of Riemann curvature tensor implying space curvature produces the body force as a space strain force. The non-zero component of Eq.(C20) is just only one equation as follows:

$$F^3 = F = E^{3330} (R_{3030} A^{30})_{:3} = E^{3330} \cdot \partial(R_{3030} A^{30}) / \partial r \quad .\tag{C21}$$

yes

I want morebooks!

Buy your books fast and straightforward online - at one of world's fastest growing online book stores! Environmentally sound due to Print-on-Demand technologies.

Buy your books online at
www.morebooks.shop

Kaufen Sie Ihre Bücher schnell und unkompliziert online – auf einer der am schnellsten wachsenden Buchhandelsplattformen weltweit! Dank Print-On-Demand umwelt- und ressourcenschonend produziert.

Bücher schneller online kaufen
www.morebooks.shop

KS OmniScriptum Publishing
Brivibas gatve 197
LV-1039 Riga, Latvia
Telefax: +371 686 204 55

info@omniscriptum.com
www.omniscriptum.com

Printed by Books on Demand GmbH, Norderstedt / Germany